Build Your Own Vacuum Form Machine

Make your own vac-u-form machine in your garage from simple hardware store

By James E. Egner II
©2007-2013. All rights reserved. ISBN 978-1-300-93157-7

Acknowledgments

A very special thanks to Thurston James. Without his excellent series of books for the prop maker, this project would never have gotten off the ground. Buy his books, available online at places like www.amazon.com . You will find that this guide is a companion to his book 'The Prop Builder's Molding & Casting Handbook".

Thanks to Blaxmyth aka Phil G. for his encouragement and mentoring. He got me started making the MDF molds, and encouraged me to try vacuum forming. Thanks to Stacey R. aka TKBIG, John, H. and Propmaster Paul for their help getting some of the details right, and for their contributions on the Replica Prop Board. Their input and suggestions really helped me be successful, and when things were not going well, they really encouraged me to continue on.

Why make this book when the information is freely available online? Well, many folks kept asking me to print them a 'shop' copy of the site. That's what this is, a 'shop' copy something you can write in, and get dirty and not have to print out.

Build Your Own Vacuum Form Machine by James E. Egner II

Contents

This project consists of several "modules" that should be built in sequence. Here is an outline:

History of my machines - Trial and error, building machines with different materials and sizes.

Module 1 - Forming surface: this is the section that will get the most "wear and tear" and is the easiest to build with common tools and materials.

Module 2 - Holding Frames: 3/4" steel tubing MiG welded. Build it sturdy, it can accommodate different plastic thicknesses

Module 3 - Forming Cart: This is made of 2x4's and a 24" square sheet of MDF. I added caster to the side so it will be easy to move this around the shop.

Module 4 - The Oven: made from a 3'x5' sheet of 1/2" Hardibacker 500 concrete backer board, a length of precoilled Nichrome wire #22, and some 60 ceramic posts. External shielding is made from the Hardibacker 500*, an angle iron 1" base is also made to hold the whole thing together. An electrical box with wire and a switch.

Module 5 - Oven Cart: Like the forming cart, this is made of 2x4's and a 24" square sheet of MDF.

Module 6 - Vacuum System: My first successfully pull using the machine was done with a 3hp shop vac. It works pretty well.

The availability of the parts is what determines the construction sequence for me. I went with the forming surface, then the holding frames, then the forming cart, then the oven, oven cart, then the vacuum system.

Safety first
To avoid any confusion, the information presented here is for personal and educational use only. If you intend to make your own machine, that's great! I hope you find this booklet helpful. However, you do so at your own risk. The author will not be held liable for any disasters, lost molds, plastics burned, blown fuse panels, or fires, burns or other accidents related to this hobby. Some common sense, please.

Have fun, and be safe.

The purpose of this booklet is to bring together in a simple printed form, the content from my web site. I keep getting asked for a printed copy of the site's content, and I understand why. It takes a lot of time and expense to print these pages out, and on a dial-up connection that can be dreadfully slow. This booklet is intended for those users.

Build Your Own Vacuum Form Machine   by James E. Egner II

History

Prototype Mark I "flip-flop" design.

Vac-u-form Table Project: Mark I Design

Prototype design:
Here is a photo album of my first attempt to build a vac-u-form, table. This is the mark I prototype, with a 12x18" heating [left side] and vac [right side]. A Sunbeam brand $25 indoor electric grill provides the heat, and a Shop-Vac brand 3 hp canister-type wet/dry shop vacuum does the sucking. Initial test prove the technology works. Based on a design found on the web, it's amazingly simple. But regulating the heat was a problem. The center of the plastic got hot first and began to sag. Drawing it closer to the heating element. Not good. I needed the hot plastic to droop away from the heating element. This called for gravity to help out.

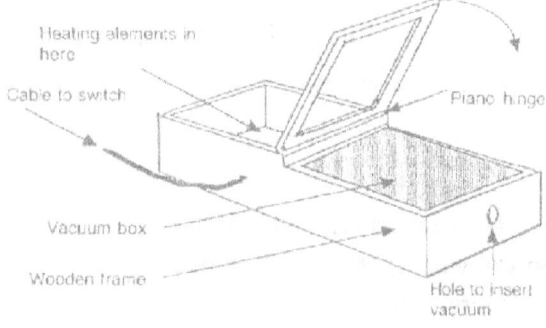

Prototype Mark I "flip-flop" design spawn from this illustration found online
http://members.aol.com/KMyersEFO/vacuum.htm%20

Build Your Own Vacuum Form Machine by James E. Egner II

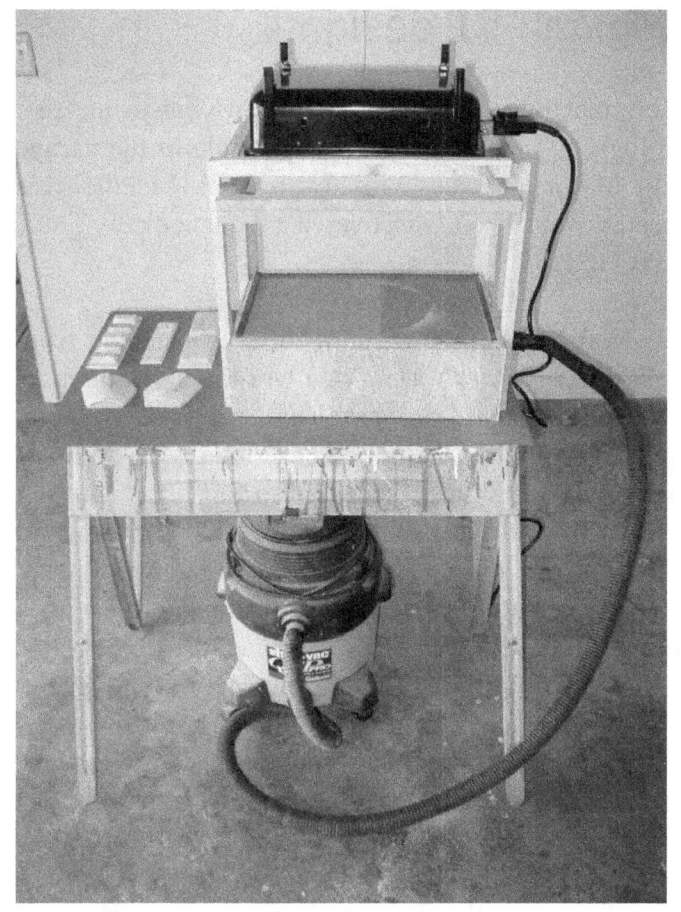

Vac-u-form Table Project: Mark II Design

The mark II had an over/under design with the heating element mounted on stilts, upside down. This should have the net effect of a more even heating effect. The weak point again is the uneven heating. A by product of the Sunbeam grill. The holding frames are held in the heating position by cabinet door magnetic catches. Much of this machine came from the Mark I designed machine. The idea for this machine came from one found online at http://www.halloweenfear.com/vacuumformintro.html

Build Your Own Vacuum Form Machine   by James E. Egner II

A Better Solution Found?

Thurston James's Vacuform machine from The Prop Builder's Molding and Casting Handbook, page 175.

The Original 'TJ' design:

My first few machines were successful. Proving to me that this technology of vacuum forming is possible in the garage or basement. My major issues with my previous machines was the poor heater/oven. Then I discovered a great book, by Thurston James.

This is the Thurston James' vacuum forming machine from his book *The Prop Builder's Molding and Casting Handbook*, page 175. This is a great book and I have most of Thurston James' books. Get them at **Amazon.com** today, if you don't have it already. His design is discussed in fair detail, but not to the extent that I could just sit down with the book and build his table. Its not a building guide. There is enough concept and general information to get started, however, and with a little time and effort, reverse-engineering and researching, this project is the way to go. There are no plans, or building diagrams in his book, except for a wiring diagram for the oven. The rest of this project has to be "re-imagined" using modern materials and substitutions, and to 'dumb-it-down' for easy construction.

A lot of vacuformers will tell you to just use pegboard, and a shop vac. This does work pretty well. The only downside is the pegboard has to be supported, and does have a tendency to sag under the weight of a large mold. The shop vac can deliver surprising results with thinner materials. But for a detailed pull on a larger machine, I wanted to follow as close as possible to what Thurston James describes. Besides this is a re-examined "practical" and cheap replica of his famous table. The following is my documentary on the production of **this** table.

Vacuum forming, 'vac-u-form', vac-forming and thermoform are interchangeable terms, and is the process of heating a thermoplastic material and shaping it in a mold with the assistance of a vacuum.

Build Your Own Vacuum Form Machine   by James E. Egner II

Module One the forming platen

Here's a look at the innards of the forming surface. Note, I used 3/4" MDF for the top and bottom and scrap 1/8" plywood for the "sandwich". I then silicone caulked the whole thing together. Make the MDF 21.5" square. Those discs you see are just some scrap material I had. Use what you have on hand. The metal is a sheet of 24x24 inch aluminum.

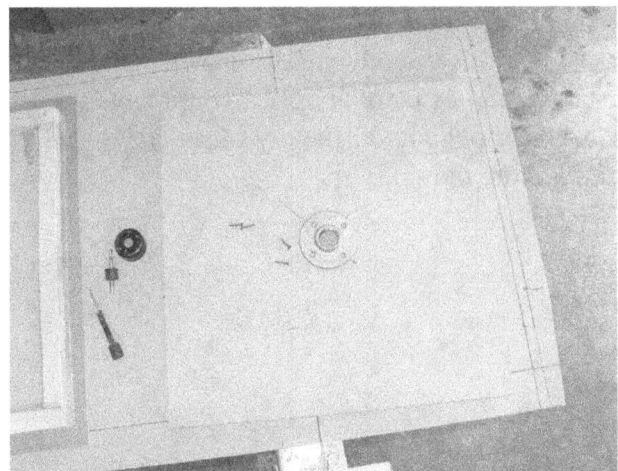

Here I am centering the floor flange. I used a 1" floor flange so that I could hook up my shop-vac right to it. For the hi-vac parts, most of the plumbing will be 3/4". See the supplies list for the needed building materials.

This is a view of the sandwich screwed together. I have positioned my aluminum sheet metal evenly under the forming sandwich. It's ready to glue down and bend. Use a good contact cement for this. Then nail and tape the corners.

Build Your Own Vacuum Form Machine

by James E. Egner II

Here is a view of the sandwich. Its clamped to the work bench, and I just use my hands to bend the aluminum sheet metal to the sides. A snip at the corners and some light tapping with a mallet and I'm done. Its ready to seal with adhesive-backed aluminum tape. I used a 1/2" staples to mount the sheet metal to the MDF.

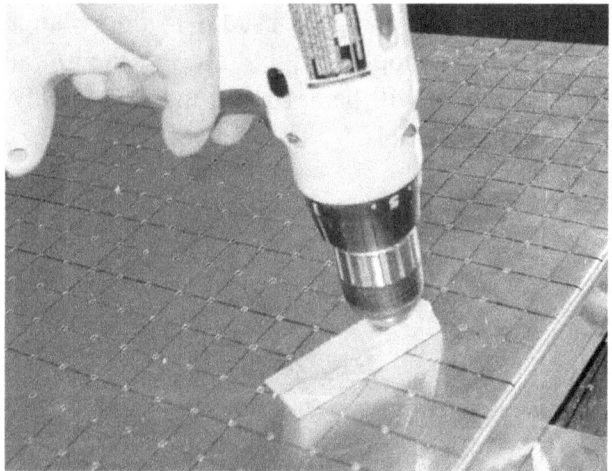

Starting from the center I drew a 1" square grid pattern. Then using the block of wood as a depth guide, I drilled the 1/8" dia. holes. Lots of holes; 484 or there abouts. I could have used a box made of 1x4s and pegboard, like my other tables. But, I wanted something more durable.

Forming surface finished! This platen is the real workhorse of the whole machine. Don't skimp on the forming platen.

Now, onto the holding frame. I already know the scale of the frame, I'm just avoiding having to break out the welding machine. Normally, the holding frame would have been made first. I found this relatively easy, skill level II.

Build Your Own Vacuum Form Machine   by James E. Egner II

Module Two the holding frames

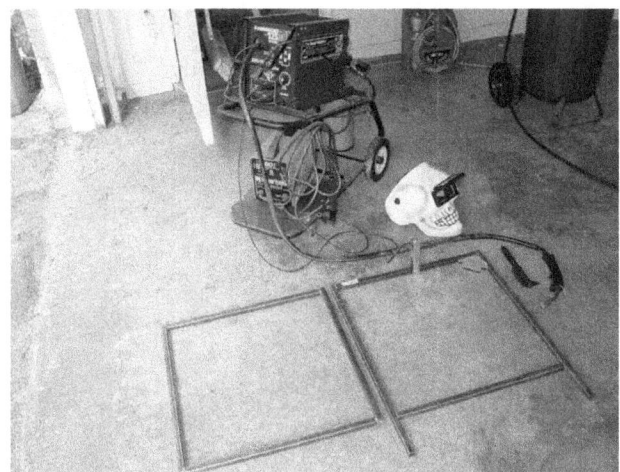

Laying out the holding frame:
Using 3/4" square metal tubing, I wanted the outside edge of the frame to be 24". Measure and cut the metal tubes. I made the bottom one with 5" extensions. Two at 24", two at 29", and four at 22.5". Try to be precise and make clean cuts. Grind down any extra metal and be sure the corners where the edges meet are plumb and flush.

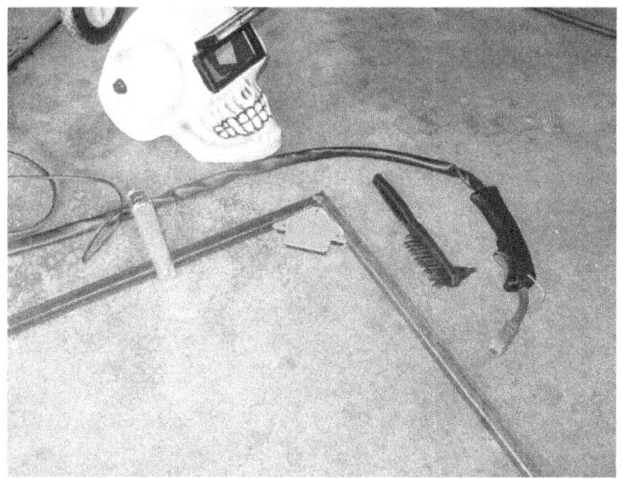

MiG weld that baby together. I used a magnetic welders guide to get the frame square. I worked on the floor mainly because I don't have a fire proof work surface. Please don't laugh at my poor welding technique. Its been 20 years since I last welded something! I really like MiG welding, and if done properly it should hold up fine. Just be sure to get a good "bead". and, yes that skull is my welders helmet.

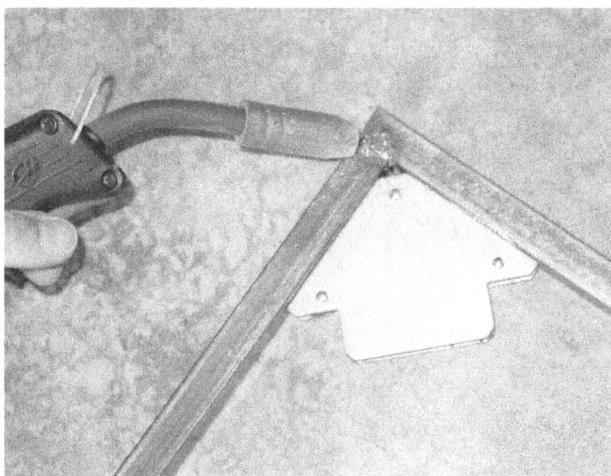

Here is a shot of the welders guide and the final weld. This process took more time to cut the tubes than to weld together. MiG welding is pretty easy, once you get the hang of it. Practice welding on some scrap first. This could have been pop-riveted together with pieces of metal straps, or brazed or stick welded. Heck, I could have made these from sturdy 1x1s and L brackets or aluminum tubes.

Note, that the frames take a pretty good beating so you will want good solid frames. I'm not sure how the aluminum or pine would hold up in the long run.

Build Your Own Vacuum Form Machine
by James E. Egner II

If you did all the work correctly, there should be a 1/2" or so space between the forming surface and the inside of the metal frame. It would look something like this. Make sure the corners of the forming surface has no sharp edges. Otherwise, it might tear the hot plastic and loose vacuum.

I found in my parts bin a bunch of these metal brackets. So I used them for the frame mount. The bolt is a 1/4" with a wing nut. I think these came from one of those toilet tank repair kits. I cut the bolt down to be just longer than the wing nut. Just use what you can, before spending money on new hardware.

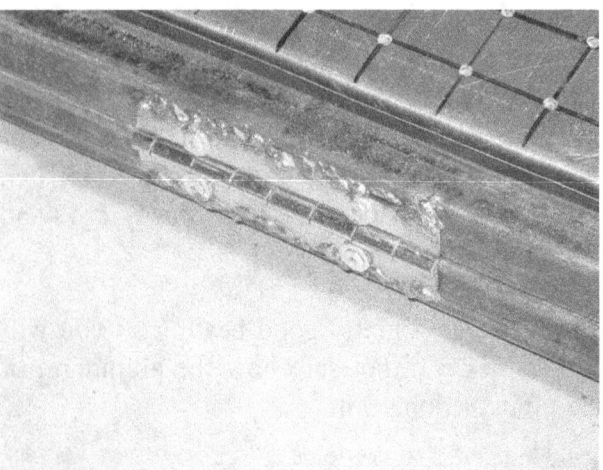

Here is some scrap "piano" hinge I cut down, and pop riveted it to the frame. I then MiG welded it to the frame. That chrome plating on the hinge sucks, so does my welding... But it gets the job done. I used a scrap piece of .080 HIPs plastic as a spacer.

Build Your Own Vacuum Form Machine   by James E. Egner II

More of those little metal brackets I used earlier for the holding frame hinge, now I'm using them to make some handles .A wooden dowel rod screwed into the holding handle. I need to make three. One on each end of the frame, the top and two sides.

The clamping fixture is a wing nut, a long eye bolt and 4 'L' brackets mounted as shown. The receiving end gets the corners ground down a bit to make it easy to tighten down. I pop riveted into position and MiG welded them to hold solid.

Here is another view of the clamping fixture. Works like a charm! Each part is easy to replace if there is an issue. I added some washers just to make it easy to operate.

Put two on the top, and one on each side, in the middle. Up next is to give the frames something to grab the plastic and hold it.

Build Your Own Vacuum Form Machine   by James E. Egner II

I then cut a sheet of MDF to the proper dimensions and bolted the frame assembly and forming surface down making a working surface. This joins all the parts together. Underneath, I cut a hole for the floor flange to poke through. This is all there is to the forming side of the vac-machine. If you just use a shop vac as your vacuum source, this becomes pretty portable.

Inside the frame I originally used some of that non skid surface tape. 3M makes the stuff for ladders and steps, and its self adhesive. It could not stand up to the heat, and soon lost it's adhesiveness. I used JBWeld, a 2-part epoxy, and some play sand. I let it dry, and it takes the heat just fine! The picture shows the 3M tape but has since been changed to the JBWeld/play sand version.

Here is a view of the forming surface with the frame in the open position for removal and installation of Styrene sheets. Really easy way to do this step. Much better than my old machines. A one-man operation to load and unload the plastic.

Build Your Own Vacuum Form Machine   by James E. Egner II

Module Three the cart

Test fitting some of the trooper armor parts. If I can pull this much at a time, that would be nice! The frame will be painted to prevent rust and to look pretty. Use the BBQ heat resistant paint. Now its off to building the cart.

Thurston James made his from square metal tubing, but said a 2x4 pine would work just as well. I wanted to keep the vacuform machine small enough to move around, so instead of making it on one big "cart" I'm putting the forming surface on one small cart, and the oven on another.

The forming cart will attach to the oven cart to make one large vacuum table. I made the cart from 2"x4" studs. The inside dimensions are 24"x24". Height is 30" to the base. The same height as my Workmate. I put casters on the side so I can tip it over and roll it around the shop. At this point the forming side is complete.

Use what you have on hand. If you don't plan to use the hi-vac system, you don't need to build the cart. You can make it a table top design.

Build Your Own Vacuum Form Machine   by James E. Egner II

Module Four the oven

Here is a shot of the Hardibacker 500* cement board material I'm using as the oven walls and floor. Basically cut a 28" square of the stuff, and then make four 28x 4.5" strips. To get the taper sides, you just measure from the center out to 24" and mark. The sides are then cut at an angle and you have the trapezoid.

This is a detail of how I attached the sides. Its a strip of aluminum with holes drilled and short screws holding it all together. Cut two of the sides 1" shorter, this allows for the thickness of the Hardibacker material. Finally to hold the sides together, I too some scrap aluminum sheet metal, and folded, trimmed it to fit, and screwed it to the sides.

The oven is sitting on the forming surface. The scale worked out fine from the book. I'll add a reinforcing walls, and a metal base soon. The walls will be made from the roof flashing material and painted black. Skill level II, but you will need a power saw to cut the concrete board.

*It has come to my attention that, over time, the high heat from the nichrome wires will cause the Hardibacker 500 to fatigue, and eventually crack. This could pose a potential fire hazard. Currently an replacement calcium silicate fiberboard is being researched. I got 3 years of regular use from my machine build with the Hardibacker 500, but anyone considering this material should understand it's limits. Consider it a less than ideal substitute.

Build Your Own Vacuum Form Machine by James E. Egner II

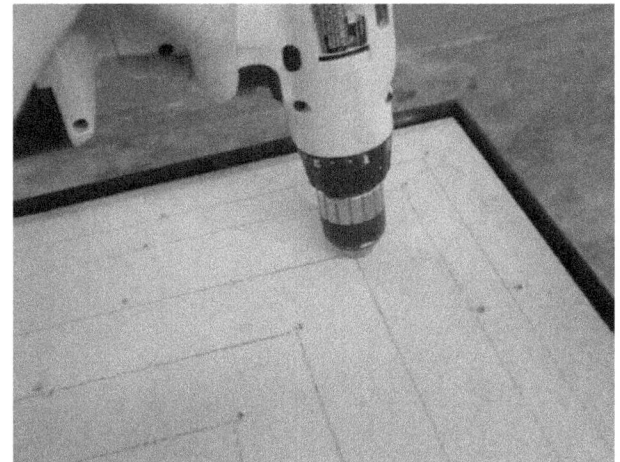

After welding a 1" angle iron material to form a sturdy base frame, I began drilling the holes for the post mounting screws. Off set all but the tapping posts 1/2 inch from the wire guide lines. Use a masonry bit that is slightly larger than the #10 screw. See the appendix for the wiring diagram and pattern. Just use a ruler, and a pencil, and transfer the diagram to the oven floor.

Here are the ceramic posts and mounting screws. I used 2" #10-24TPI for the terminal posts and a top mounted wing nut. These ceramic posts and the nichrome wire can be purchased from http://www.infraredheaters.com/nicrcoil.htm. See the supply list for quantity. The posts and wire are the only unique material that might not be available at your local hardware store.

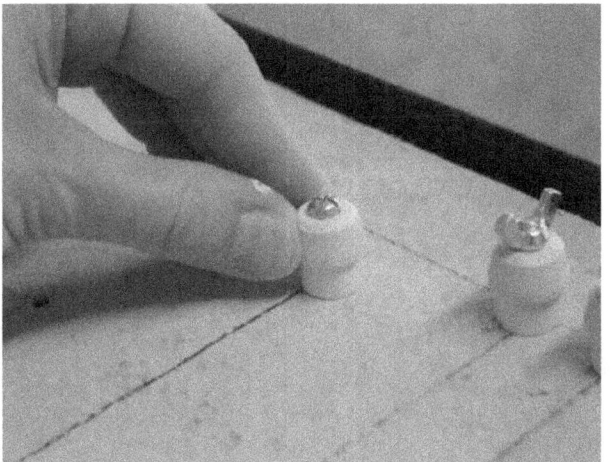

Some test fitting.. Regular ceramic post and terminal post [the one with the wing nut]

Build Your Own Vacuum Form Machine   by James E. Egner II

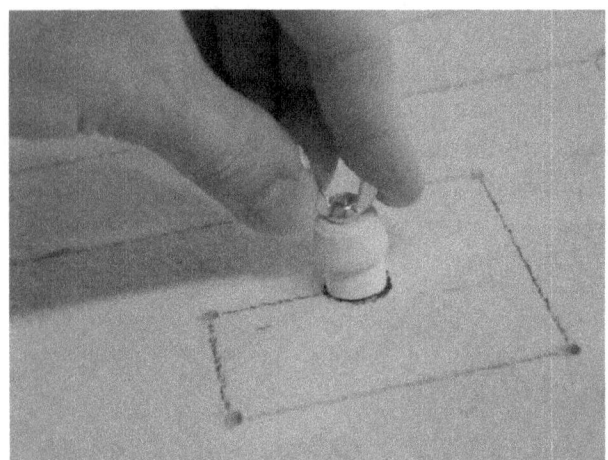

Starting from the center... Mount the screws. The underside of the backer board might need some scraping, as the holes drilled through are rough on the underside. Remember to mount the tapping points on the lines, and offset the posts from the line.

And this is what mine looked like after drilling and mounting all 60 of the ceramic posts and terminal posts. Note the metal frame, and the cart! The metal frame is just 3/4-inch angle iron welded.

Here is a detail shot of the way I mounted the oven frame to the cart.

Build Your Own Vacuum Form Machine   by James E. Egner II

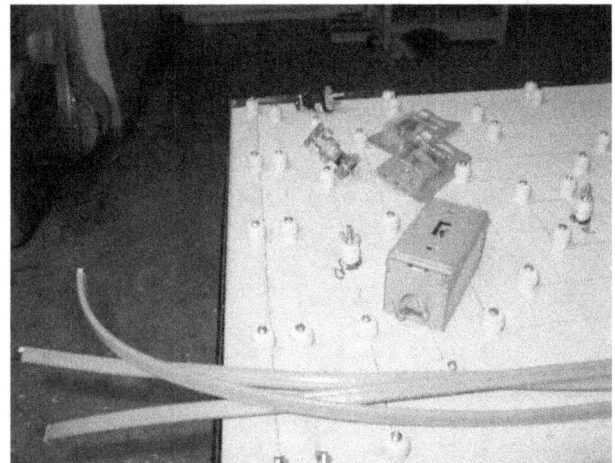

Now for the wiring. I started with the high voltage first. Using 10-2G wire, some real big wire nuts, a 20 AMP switch, an electrical box and cover, and a plug. See the Wiring section for details.

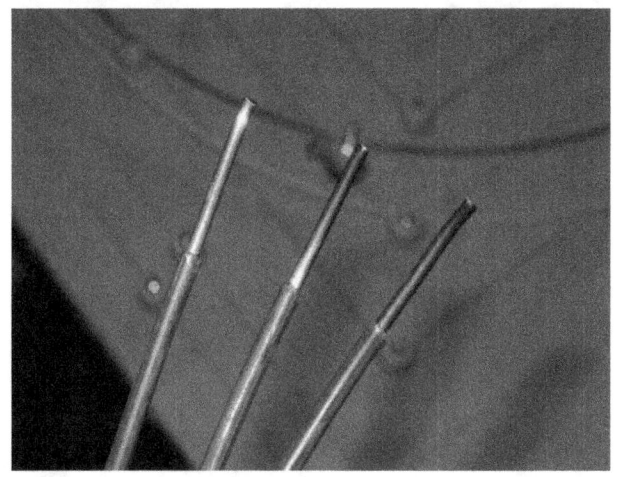

Cut three, 3-foot segment of wire, and stripped off the orange outer cover. Then removed the black, white and bare wires. Then strip the ends off the black and white wires like so. These wires will be wrapped around a 2" bolt, so don't strip too much wire!

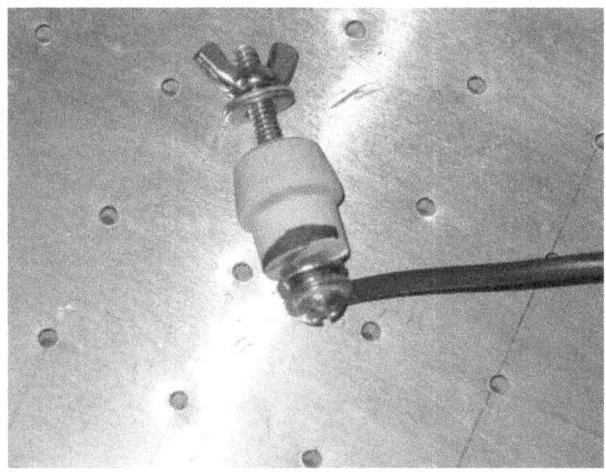

This is test fitting of the terminal post. Starting from the bottom, its a 2-inch #10-2 bolt, #10 washer, Wire bent to form around the bolt, another washer, a #10-24 nut, tightened up, another washer, the ceramic spacer, two washers on top and finally a #10-24 wing nut.

Build Your Own Vacuum Form Machine by James E. Egner II

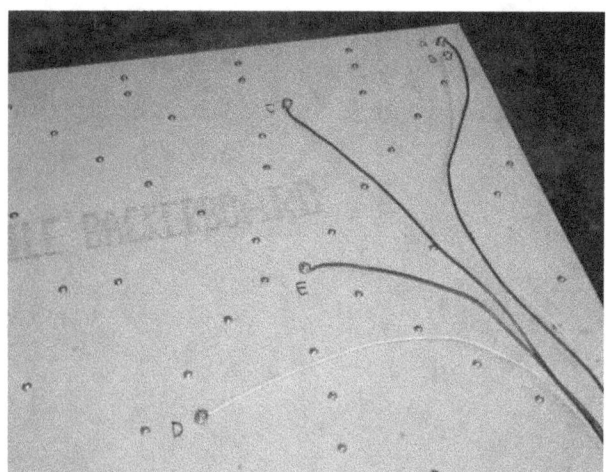

Now for the wiring. I started with the high voltage first. Using 10-2G wire, some real big wire nuts, a 20 AMP switch, an electrical box and cover, and a plug. See the Wiring section for details.

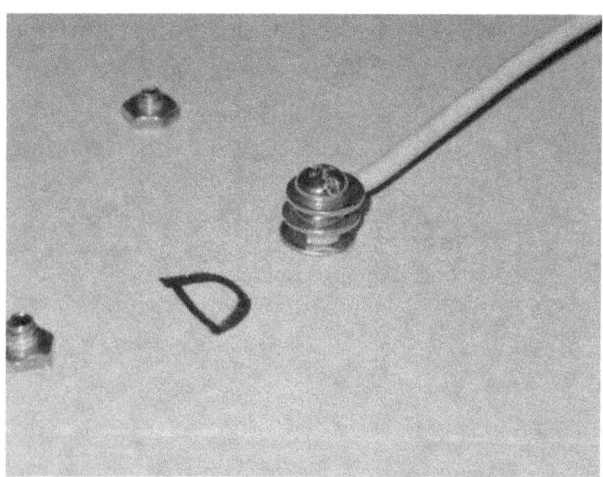

Cut three, 3-foot segment of wire, and stripped off the orange outer cover. Then removed the black, white and bare wires. Then strip the ends off the black and white wires like so. These wires will be wrapped around a 2" bolt, so don't strip too much wire!

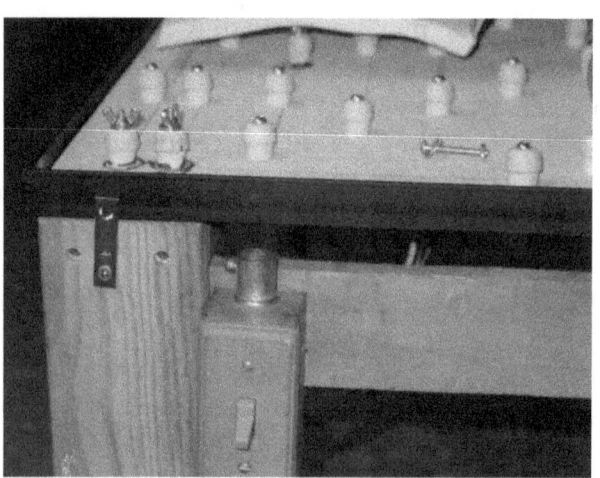

This is test fitting of the terminal post. Starting from the bottom, its a 2-inch #10-2 bolt, #10 washer, Wire bent to form around the bolt, another washer, a #10-24 nut, tightened up, another washer, the ceramic spacer, two washers on top and finally a #10-24 wing nut.

Build Your Own Vacuum Form Machine **by James E. Egner II**

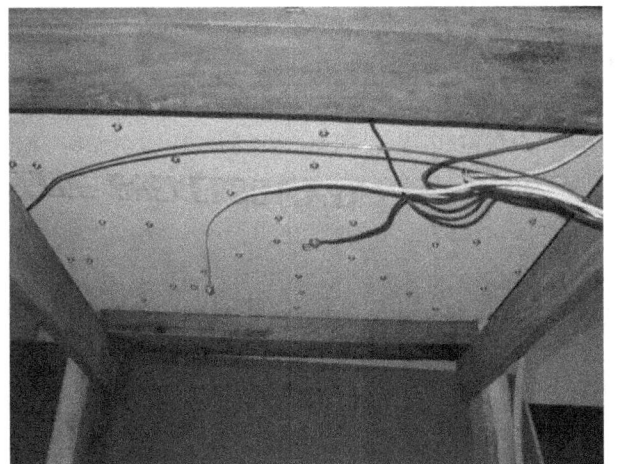

Wire pulled to the switch box. All that is left to do is wire these to the switch, add a power cord, and I'm ready to add the heater coils. Wiring a light switch basically means take all the black wires and tie them together with a 4" pigtail using a wire nut. Then wire all the white wires, but this time include the white wire from the power cord. The switch has the black wires connected only. The black wire from the power cord, and the black wire from the pigtail.

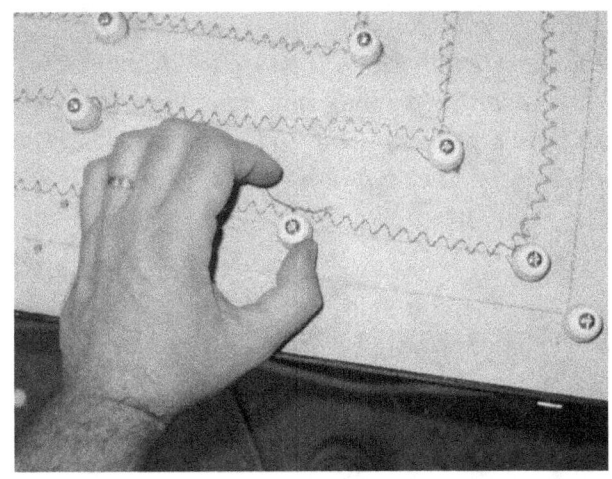

I got excited and forgot to photograph the wire stretching, so all I have photo-wise is a shot of the small wires I used to tie off the heater coils. This process was easy, but my biggest concern was where to start. The instructions in the book, call to start from the center of the segment, and start from post D. The wires have a lot of slack in them, and I'm not sure if that's good or bad.

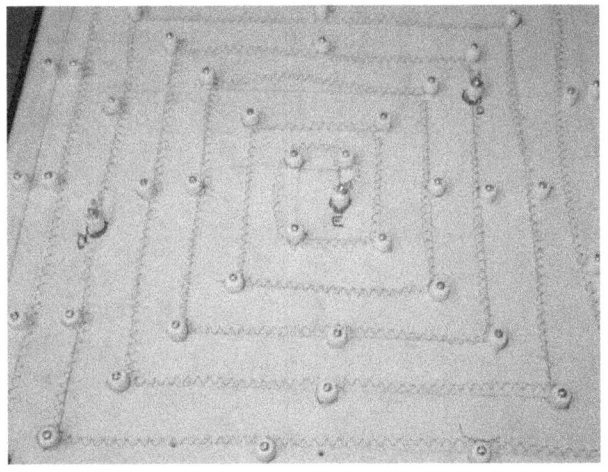

And the finished oven coils all tied down and ready to test. Kind of saggy coils, but that's OK. After a few test, the last thing to do is add some heat reflectors on the walls to better control the heat flow. The corners are going to be the cold spots.

Build Your Own Vacuum Form Machine
by James E. Egner II

Here is the finished oven sans the lid and frame. The next challenge will be to test the rig without blowing a fuse or breaker! I also want to try to convert this to a table top machine and ditch the carts. Skill level III, mainly because the oven parts are hard to get, and the safety concerns regarding the proper wiring.

I have been asked to take a Ohms reading. Here are the results: each segment draws about 5 ohms. The resistance varies from 4.9 to 5.1 ohms. I assume this is pretty good considering the home-made nature of the oven. Total of 20.1 ohms across the whole oven coil.

*It has come to my attention that, over time, the high heat from the nichrome wires will cause the Hardibacker 500 to fatigue, and eventually crack. This could pose a potential fire hazard. Currently an replacement calcium silicate fiberboard is being researched. I got 3 years of regular use from my machine built with the Hardibacker 500, but anyone considering this material should understand it's limits. Consider it a less than ideal substitute.

Appendix: Oven coils

From John H. regarding the wire you get from www.infraredheaters.com. [The precoiled type vs. the instructions found in Thurston James' book]

"Well, correct me if I'm wrong, but the book says to wrap the wire around a 1/4" rod. Infraredheaters.com pre-coiled wire has an approximate inside diameter of 1/4" so, I think it's the right size as far as that is concerned.
So basically to calculate how much pre-coiled wire I would need, I first figured out how many turns per inch the 22 gauge wire gave me by taking the outside diameter subtracting the inside diameter and dividing that by 2 (.30 - .25)/2 = .025 wire thickness. I then took 1" and divided that by .025 and got 40 turns per inch.

Then I took 88' converted it into inches (88*12) then divided it by 40 (26.4) then converted it back into inches (26.4/12) and came up with 2.2 feet of coiled wire to do the entire job. If you look at the pic on page 184 it shows the guy holding 1/2 of the wire required for the entire job and it looks to be about a foot or so. So I think this calculation is correct.

With that said, we should both be able to split the cost of the 10' pre-coiled wire and still have plenty left over. "

And for the info regarding how much to stretch:

First I cut a 2' 1/4" [24.25 inches] length from the 5 feet of coiled wire, then cut it in half to make two 1' 1/8" [12.125 inches] lengths. I then marked the center of each length of wire at 6 1/16" with a Sharpie Marker. After that was done, I put the marked center on a nail, grabbed both ends and stretched the wire to 95 1/2".

WARNING: Use a pair of pliers to hold the ends! I made the mistake of stretching one of the pieces with my bare fingers and wound up with a nice puncture wound under my finger nail... ouch!

Thanks John!

Build Your Own Vacuum Form Machine   by James E. Egner II

Appendix II: Oven wiring

Supplies:

10 gauge wire: It will be called "Romex 10-2 G" in the stores most likely. It means 10 gage, 2 conductors with Ground [I bought a 25 foot length]
Dimmer switch: Preferably one with a distinct on/off so you can be sure your oven's off. If you can find one That lights up when it's turned on, that would be perfect. [I used a gray 20 amp light switch]
Electrical box and cover: Preferably metal and wide enough for one light switch (get the deepest one you can find so you have some room to work). Some boxes have a metal piece that can tighten down on the wires coming in so the connections don't get pulled on. Get this kind if you can. Also buy a cover for it, which will probably be plastic. You want one that fits your dimmer switch.
"Wire nuts": They come in different sizes, so realize that you need two: one to join (4) 10 gage wires and one for (3) 10 gage wires. The box will list how many of each gage of wire you can join together with it. (#14, 2-5 #12, 2-3 #10, 2 or something like that). You may need to buy more than one kind, but probably not. You can usually buy small packages (10 or 20) of these instead of the big boxes for electricians.
Electricians tape.
3 Prong plug: With nothing attached to it if you can. Get one rated at 30 amps. If you can't find one, buy a heavy duty extension cord and cut the end off that you don't need. [I used a 3 prong heavy duty 15 amp plug]

Construction:

Here's what you do:
Cut off 3 pieces of wire, 2-3 feet long. Then remove the outer covering. Inside you'll find a black wire (HOT), a white wire (NEUTRAL), and either a green or a bare wire (GROUND).
Strip off the last 1/2 inch of insulation on the 3 black wires with either a knife (just cut all the way around and slide the insulation off, or with a wire cutter being careful to only cut the insulation, not the wire)
Connect BLACK wires to the A,C,E terminals from the diagram. Wrap the bare wire around the bolt clockwise, then tighten the nut down onto it. Then cover with electricians tape so no metal is exposed.
Connect WHITE wires to the B,D terminals the same way.
Mount the box to your oven with screws. Put it in a place that's convenient for your switch.
Connect the ground wire to something metal in the oven (not the coils).
Thread all the wires into your box through a side hole. Tighten the metal piece down on the wires, and trim off the excess wire, leaving 4 or so inches hanging out of the box.
Strip the ends of your wires.
Cut a 4 inch black wire, strip the ends, and make a bundle with the ends of your 4 black wires so the ends point in the same direction. Twist the ends together as best you can since they're thick. Twist the ends clockwise. Put a wire nut on the end and screw it on clockwise with your hand until it won't go any further. Make sure all wires are secure.
Do the same with your 2 white wires, but instead of using a short wire, connect the 2 to the white wire from your power cord.
Connect the short black wire to your dimmer switch using the screw terminals on the side.
Connect the black wire from your plug to one side of your dimmer switch.
Connect the ground wires from the oven and the power cord to the metal box.
Stuff everything into the box. Screw the switch to the box, and screw on the cover.

Build Your Own Vacuum Form Machine   by James E. Egner II

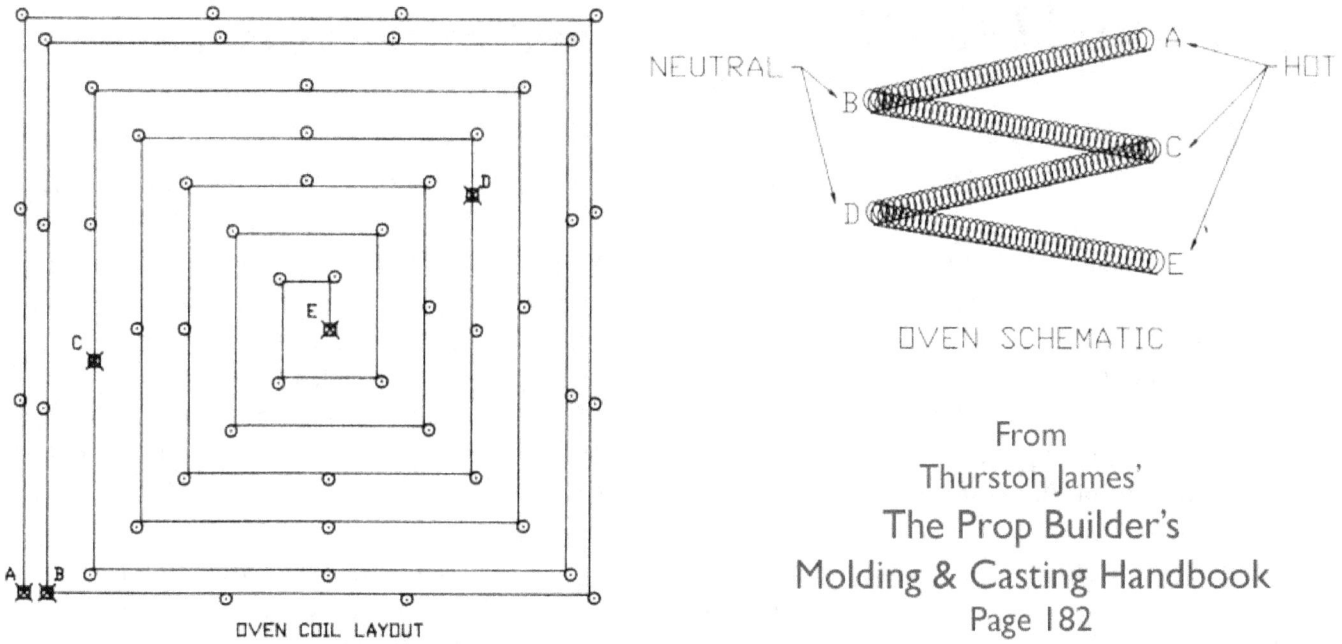

Figure 1. Terminal post and wiring diagram for 24x24 inch oven design.

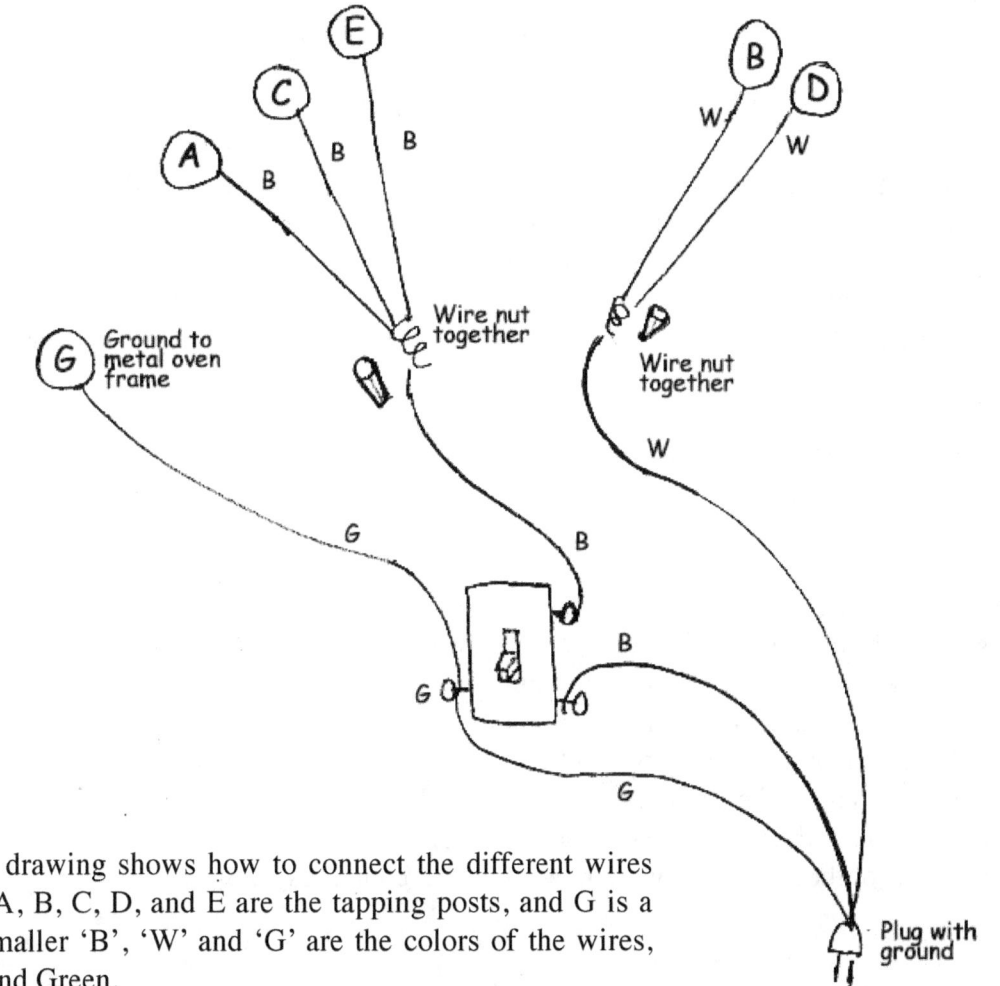

Figure 2. This drawing shows how to connect the different wires to the switch. A, B, C, D, and E are the tapping posts, and G is a ground. The smaller 'B', 'W' and 'G' are the colors of the wires, Black, White and Green.

Build Your Own Vacuum Form Machine   by James E. Egner II

Appendix II: materials, and vacuum details

Oven Materials
(Calcium silicate is expensive!) A working substitute for the oven is James Hardi's Hardibacker 500*. Available from home improvement centers. Look in the floor tile or bathroom shower sections of the store. A 3'x 5' x 1/2" sheet is $10. They are basically a concrete board 60% calcium silicate. I've talked with a chemical engineer about the plausible use of this and he seems to think it will work, but I need to use a 1/2" inch thick sheet instead of the 3/8". It will get hotter and might not insulate the heat as well. Also, the engineer suggested lining the interior of the oven with sheet metal to help support the material. The first test of the oven has been successful, and the Hardibacker 500* seems to work fine. Including the angle bracket and screws, the $60 for the ceramic posts, and another $10 for the NiChrome wire, and factor in another $15 for the electrical, the cost of the oven is around $95.00.

It has come to my attention that, over time, the high heat from the nichrome wires will cause the Hardibacker 500 to fatigue, and eventually crack. This could pose a potential fire hazard. Currently an replacement calcium silicate fiberboard is being researched. I got 3 years of regular use from my machine built with the Hardibacker 500, but anyone considering this material should understand it's limits. Consider it a less than ideal substitute.

Vacuum system
The proper tank to use for a hi-vac system is a $144 narrow 30 gallon hot water heater tank from the home improvement center. For my budget, its just too expensive. My test pull was done with a 3hp Shop-Vac, and it seemed to work fine. Cost = $0.00 since I've had the shop-vac for 10 years!

There are two alternate ways to improve the vacuum. One will be to use a second shop-vac. I can use two shop-vacs in sequence and get more in/Hg instead of the 3-4 in/hg from a single unit. Also, a check valve might be a good idea, to help hold the vacuum once it's pulled. I've also purchased a sump pump check valve, that can be modified to use a spring loaded check valve, and a hi-vac low volume source. I'm considering my options here. From the book "Do It Yourself Vacuum Forming for the Hobbyist" by Douglas E. Walsh.

For the pump/tank vac system, Gast sells different models and the one Thurston James suggests is Model 1065. Its actually a 8.5 CFM unit that requires a motor. Gast Model 1023 is a 10 CFM unit with a built in motor.

Vacuum Pump with motor
Check valve [so the system does not back-draw any air through the vacuum pump]
Air Filter [so there are not particles to destroy the pump]
Tank [30 gallon, 15 is too small; not enough volume]
Pressure switch [to keep from imploding the tank]
Vacuum gauge [so you can see the amount of vacuum and the vacuum rate]
Valves [to close off parts of the system and to open the table valve to start the pull]
Plumbing fittings [lots of these, nipples and elbows, flanges, couplings, etc.]

Build Your Own Vacuum Form Machine by James E. Egner II

Appendix III: supplies etc.

Forming Supplies:
Locally, no one seemed to know what the stuff is. After some research I came to the conclusion that Polystyrene is a chemical name, High-Impact Styrene Sheets is what you want to search for. Sometimes called HIPs [High-Impact Polystyrene Sheet] I've had the best luck searching for HIS [High-Impact Styrene Sheet] Below are some suggested sources. Be sure to get 48"x96" sheets not the 40"x72" sheets. You will need between .060 and .080 white. No texture. This material will have to be gloss coated to get that shiny appearance. A harder to form material called ABS is used by some, but requires a hi-vac system for tight pulls.

www.ProfessionalPlastics.com [good customer service. Thanks Lydia!]

www.USplastic.com [claims they sell the .080 in 48x96 inch sheets, but I can't find it in their catalog. Only the smaller 40x72 inch sheets.]

Basic Math:
Some basic conversions:
.125 is 1/8th inch thick
.062 is 1/16th inch thick
.080 is 1/12th inch
.093 is 1/11th

Ohms:
I have been asked to take a Ohms reading. Here are the results: each segment draws about 5 ohms. The resistance varies from 4.9 to 5.1 ohms. I assume this is pretty good considering the home-made nature of the oven. Total of 20.1 ohms across the whole oven coil.

Build Your Own Vacuum Form Machine   by James E. Egner II

Appendix IV: operation details

I managed to make several pulls using my new contraption. Here is what I learned:

Make a lid. I used the leftover Hardi-backer 500 material that I had from the oven construction, and it makes a nice lid. I made a handle from some angle brackets and a piece of dowel rod and some large washers. Works great. It speeds the oven heating, and plastic melt time. But watch out! You can over heat the plastic! I did! Man, what a mess, and smell!

I added a second handle to the frame. This gives extra pressure to hold the frame to the molds and forming surface. You will need the extra downward pressure to get a good vac seal.

The oven needs time to heat up. You can start pulling hot plastic in say 5 mins. But the longer time the oven has to heat up, the better the quality of pull you get. With the oven hot, the corners even get nice and soft! That's convection heating working for you. Don't rush the oven preheat.

Keep the lid on at all times. Even when you are heating the plastic. It sits right on the frame and helps hold the heat in. Set it aside only when you are pulling the hot plastic. Have an assistant replace the lid on the oven as soon as possible to hold all that hot air in.

Wear leather gloves. When you start to pull a mold, you can massage the hot plastic to get into the stubborn areas. Also, it keeps you from burning your hands!

I used HIS [High Impact styrene sheet] in a .080 thickness. It works great, but to be honest, on small parts, you really don't need that thick of a piece of plastic. It does however work great for the larger [deeper pulls] pieces.

The molds all need risers. Period. Otherwise you get a curved bottom, and that means poor fitting parts. Also, the molds need to be fine tuned for easy release from the melted plastic. A slight taper, grooves, or something to help the parts slip out from the vac-ed plastic. Each mold had to be pounded out! I even broke a mold trying to release it. The slightest undercut can cause this.

The imperfections on my molds even show up on the .080 thick material. I thought it would not be noticeable. I was wrong. Time to break out the Bondo and sanding paper!

Forming time is pretty short, say 15 8-10 seconds, with another 5 seconds that you can massage the plastic with. After that, the plastic is so still it won't bend anymore. It takes a good 30 seconds to cool to touch enough to remove the plastic from the frames. If the molds will release better, the production could go pretty fast. Right now, I have to pound the molds from the plastic.

Overall I pulled seven panels in about one hour. 2 were good enough to give away to a friend who wanted the parts, one was a boo-boo, total melt-down, and one was not hot enough to form. One was OK and I trimmed out the parts, and one was decent, but not good enough to use any of the parts. It does make an interesting wall art decoration though!

Build Your Own Vacuum Form Machine   by James E. Egner II

Appendix V: supplies

QTY.	Name	Size	Source
1 Sheet	MDF [Medium Density Fiberboard]	3/4" by 4' by 8'	Lowes or Home Depot
3 tubes	square metal tubing	3/4" by 3/4" by 6'	Lowes or Home Depot
6	#2 2x4 pine or fir	stud length	Lowes or Home Depot
1	3/4" Floor flange or 1" floor flange	3/4" or 1"	Lowes or Home Depot
1 sheet	Sheet metal [aluminum]	24" by 24"	Lowes or Home Depot
1 tube	silicone caulk		Lowes or Home Depot
2	piano hinge hardware	3/4" x 4"	Lowes or Home Depot
2	bolts for clamping hardware		Lowes or Home Depot
2	wingnuts		Lowes or Home Depot
1	dowel rod	3/8" dia.by 5"	Lowes or Home Depot
1 sheet	Hardibacker 500 backer board**	1/2" x3'x5'	Lowes or Hoime Depot
1 roll	nichrome wire #22,	about 5 feet pre-coiled	Infraredheaters.com
56	ceramic posts item # PI-B-1-S		Infraredheaters.com
1	Vacuum Pump 10 CFM Gast model 1065, 1023 or substitute *		10 CFM Gast, Ebay
3	3/4" ball type valves *	3/4" threaded	Lowes or Home Depot
1	hot water tank *	30 gallon "slim" hot water	Lowes, or Home depot.
1-2	3hp or better Shop-Vac	3hp or better	Lowes or Home Depot
1	Sump Pump check valve	1-1/4 to 3/4"	Lowes or Home Depot
1	10-2-G electrical wire	25'	Lowes or Home Depot
1	20am light switch [gray]	20 amp with ground	Lowes or Home Depot
2	large wire nuts#10 - 4 wires		Lowes or Home Depot
1	Light switch box with cover	std	Lowes or Home Depot
1	3 prong electrical plug with ground	20 amp	Lowes or Home Depot
1 box	1-1/2 inch screws with nuts [50]	#10-24	Lowes or Home Depot
1 bag	2 inch screws [5]	#10-24	Lowes or Home Depot
1 bag	flat washers [25]	#10	Lowes or Home Depot
1 bag	wing nuts	#10	Lowes or Home Depot

    * not used so far in this project

    ****It has come to my attention that, over time, the high heat from the nichrome wires will cause the Hardibacker 500 to fatigue, and eventually crack. This could pose a potential fire hazard. Currently an replacement calcium silicate fiberboard is being researched. I got 3 years of regular use from my machine build with the Hardibacker 500, but anyone considering this material should understand it's limits. Consider it a less than ideal substitute.*

This is not a complete list of materials. Use what you can find on hand, as I did. This can save you a lot of time and money.

Conclusion

Whoa! It worked! It really worked. I'm just blown away. Why commercial vacuform/thermoform machines are so expensive is beyond me. The hardest part by far to make is the oven, and only because of the exotic nichrome wire. calcium silicate fiberboard, and the ceramic posts. If there are hardware store substitutes for these items, please let me know. Every hobbyist should make a vacuform machine. If your into model railroad, RC car and airplanes, scale models, arts and crafts, prop replicas, costumes, or any other hobby, get to work on your vacuform machine NOW! Its that easy. Total cost for the test pull: Less than $150.00. The molds are more work, but once you make a good hearty mold, you can pull as many copies as you want.

It has come to my attention that, over time, the high heat from the nichrome wires will cause the Hardibacker 500 to fatigue, and eventually crack. This could pose a potential fire hazard. Currently an replacement calcium silicate fiberboard is being researched. I got 3 years of regular use from my machine build with the Hardibacker 500, but anyone considering this material should understand it's limits. Consider it a less than ideal substitute.

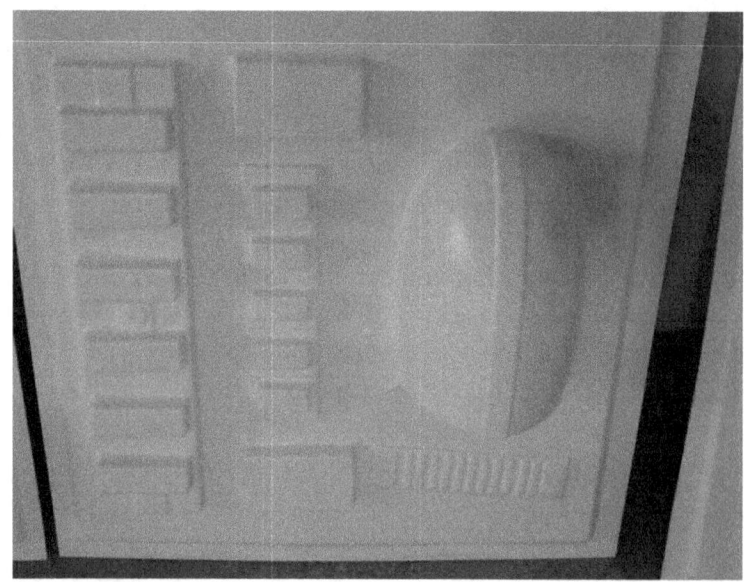

References

http://members.aol.com/KMyersEFO/vacuum.htm%20

The Prop Builder's Molding and Casting Handbook, page 175. Thurston James, Better Way Books.

The Prop Builder's Molding and Casting Handbook, page 182. Thurston James, Better Way Books.

Do It Yourself Vacuum Forming for the Hobbyist by Douglas E. Walsh.

http://www.tk560.com/phpBB2/index.php - discussion board with additional information regarding DIY vacuum forming machines.

Additional Information

Please visit http://www.tk560.com for additional information regarding this project. Also, be sure to sign up for the discussion board. There are a lot of talented folks that contribute to the vacuum forming section. http://www.tk560.com/phpBB2/index.php

Build Your Own Vacuum Form Machine by James E. Egner II

A Note on Safety

*It has come to my attention that, over time, the high heat from the nichrome wires will cause the Hardibacker 500 to fatigue, and eventually crack. This could pose a potential fire hazard. Currently an replacement calcium silicate fiberboard is being researched. I got 3 years of regular use from my machine build with the Hardibacker 500, but anyone considering this material should understand it's limits. Consider it a less than ideal substitute.

Build Your Own Vacuum Form Machine

Make your own vac-u-form machine in your garage from simple hardware store

By James E. Egner II
©2007-2013. All rights reserved. ISBN 978-1-300-93157-7

www.ingramcontent.com/pod-product-compliance
Lightning Source LLC
Chambersburg PA
CBHW081147170526
45158CB00009BA/2758